Space Capitalism

The Trillion-Dollar Race Beyond Earth

Mike Bhangu

BBP Copyright 2025

Copyright © 2025 by Mike Bhangu.

This book is licensed and is being offered for your personal enjoyment only. It is prohibited for this book to be re-sold, shared and/or to be given away to other people. If you would like to provide and/or share this book with someone else, please purchase an additional copy. If you did not personally purchase this book for your own personal enjoyment and are reading it, please respect the hard work of this author and purchase a copy for yourself.

All rights reserved. No part of this book may be used or reproduced or transmitted in any manner whatsoever without written permission from the author, except for the inclusion of brief quotations in reviews, articles, and recommendations. Thank you for honoring this.

Published by BB Productions

British Columbia, Canada

thinkingmanmike@gmail.com

Space Capitalism
The Trillion-Dollar Race Beyond Earth

Table of Contents

Introduction: Welcome to the Final Frontier

Chapter 1: From Sputnik to SpaceX — How Capitalism Hijacked the Final Frontier

Chapter 2: Asteroid Mining — The Ultimate Get-Rich-Quick Scheme (Until Space Pirates Show Up)

Chapter 3: Lunar Real Estate — Buy Your Acre of Moon Dust Today! (Terms and Conditions Apply)

Chapter 4: Interplanetary Trade — Why Your Amazon Package Will Take 300 Years (and Cost a Planet)

Chapter 5: Space Tourism — Where Billionaires Go to Escape Their Tax Bills

Chapter 6: The Dark Side of Space Capitalism — Pollution, Plutocrats, and the Probability of War Over Uranus

Space Capitalism

Introduction: Welcome to the Final Frontier

The Universe is Open for Business (And Yes, Bezos is the Manager)
Once upon a time, space was about heroes in foil suits planting flags and eating Tang. Today? It's about billionaires in rocket-shaped midlife crises outbidding each other for the best parking spot on Mars. Welcome to Space Capitalism: The Trillion-Dollar Race Beyond Earth—your guide to the greatest show in the galaxy, where the stakes are astronomical, and the business plans are... well, let's just say Elon Musk once funded a rocket by selling flamethrowers.

This book isn't about whether we'll colonize the cosmos. It's about how we'll turn it into a combo platter of Silicon Valley hubris, interplanetary tax evasion, and Yelp reviews for asteroid mining companies. Buckle up, Earthling. The future is a circus, and the clowns have PhDs in astrophysics.

From Sputnik to Space Junk: A Brief History of Human Shenanigans
In 1969, Neil Armstrong took a "giant leap for mankind." In 2023, Jeff Bezos took a giant leap for his LinkedIn profile, floating in zero-G while Amazon workers union-busted in the background. How did we get here? Let's recap:
- 1960s: "We choose to go to the moon!" Translation: "We choose to spend 4% of the U.S. GDP to dunk on the Soviets."
- 2000s: "We choose to monetize the moon!" Translation: "We choose to sell lunar timeshares to people who still lease their iPhones."

The Cold War was a simpler time. Back then, we feared nuclear annihilation. Now? We fear Elon's Twitter feed. Progress!

Meet the Cast: The Rat Pack of Rocket Science

No tale of cosmic capitalism is complete without its protagonists:

1. Elon Musk (Tony Stark's Chaos Gremlin Cousin):
   - Claims he'll die on Mars. Not because it's noble, but because he'll forget to pack oxygen.
   - Achievements: Reusable rockets, Starlink satellites, and convincing people to care about Dogecoin.

2. Jeff Bezos (The Walmart of the Milky Way):
   - Blue Origin's motto: "Gradatim Ferociter" (Latin for "Slow and Steady Wins the Race to the Edge of Space for 11 Minutes").
   - Currently auctioning naming rights to Jupiter's storms. Hurricane Prime coming soon.

3. Richard Branson (Space's Drunk Uncle):
   - Showed up to the space race with a joystick, a bottle of bubbly, and a Virgin Galactic logo plastered on everything.
   - Offers frequent flyer miles for suborbital joyrides. "Collect 10 and get a free oxygen tank!"

Together, they're the Horsemen of the Space Apocalypse, here to sell you a timeshare on Europa.

What You'll Learn (Besides How to Cry in Zero-G)

This book is your all-access pass to the dumpster fire we're launching into orbit. You'll explore:
- Chapter 1: How NASA became SpaceX's Uber driver.
- Chapter 3: Why your moon deed is worth less than a Chuck E. Cheese token.
- Chapter 5: The art of vomiting elegantly during a $50 million space joyride.
- Chapter 6: Why war over Uranus is inevitable (and grammatically confusing).

You'll laugh. You'll cry. You'll question why you ever donated to that "Save the Earth" fundraiser.

Why Should You Care? (Spoiler: You're Probably Already Bankrolling This)

Let's be real: You're not rich enough to buy a ticket to space. But you are rich enough to fund it! Through taxes, tech stocks, and that $400 you spent on a Stanley cup, you're propping up the greatest grift in galactic history. This book answers the big questions:
- Will space capitalism save humanity?
- Or just give us a new place to ruin?
- And why does Jeff Bezos's rocket look like... that?

By the time you finish this book, you'll either:
1. Be inspired to launch your own startup (Asteroid Tinder: Mining Love in the Stars!).
2. Develop a crippling fear of the words "Mars colony."
3. Realize your grandkids will inherit a planet-sized student loan for their moon condo.

Chapter 1: From Sputnik to SpaceX — How Capitalism Hijacked the Final Frontier

The Day Space Got a Price Tag

Let's start with a bang. Literally. On October 4, 1957, the Soviet Union launched Sputnik 1, a beach-ball-sized satellite that beeped ominously as it orbited Earth, terrifying Americans in a way only a communist space beep could. The U.S. response? Panic, followed by a lot of taxpayer-funded rocket science. Thus began the Cold War Space Race: a dick-measuring contest between superpowers where the ruler was a missile, and the prize was bragging rights to the moon.

Fast-forward to 2023. The biggest rivalry in space isn't USA vs. Russia—it's Elon Musk vs. Jeff Bezos in a passive-aggressive Twitter feud over who can build the fanciest space yacht. How did we go from "One small step for man" to "One giant leap for my IPO"?

The Cold War: When Space Was a Patriotic Hobby (and Also Kinda Free)

Back in the 1960s, space was a bipartisan vanity project. NASA's budget ballooned faster than a Saturn V rocket, fueled by JFK's iconic "We choose to go to the moon" speech. (Spoiler: He never mentioned how we'd pay for it.) For context, the Apollo program cost about $288 billion in today's money. That's roughly $1,000 per American… or as Bezos calls it, "loose change in the couch of my superyacht."

But here's the kicker: Nobody expected a ROI. The goal wasn't profit—it was propaganda. Astronauts were heroes, rockets were symbols, and the moon landing was basically America's version of a TikTok flex: expensive, dangerous, and absolutely worth it for the clout.

Enter the "Billionaire Space Bros" (Or: How to Gentrify the Galaxy)
Cut to the 21st century. The Cold War ended, NASA's budget got downsized to "public radio pledge drive" levels, and suddenly, space was up for grabs. Cue the Silicon Valley billionaires, who looked at the stars and thought: "Hmm, untapped market."

- Elon Musk (SpaceX): The Tony Stark of South Africa-by-way-of-Twitter. Founded SpaceX in 2002 with the modest goal of making humans a multiplanetary species (and also selling flamethrowers). Achievements include: reusable rockets, Starlink satellites that outnumber pigeons, and tweeting memes about Dogecoin on Mars.

- Jeff Bezos (Blue Origin): The guy who invented next-day delivery and then decided to deliver himself to space. Blue Origin's motto: "Gradatim Ferociter" (Latin for "Slow and Steady Gets You to the Edge of Space for 4 Minutes").

- Richard Branson (Virgin Galactic): The British uncle who shows up to the space party with a suborbital joyride and a bottle of champagne. His pitch: "Why colonize Mars when you can take a selfie in zero-G?"

Together, they're like the Rat Pack of Rocket Science—if Frank Sinatra had ever sued Dean Martin over lunar mineral rights.

NASA's New Hustle: "We're Not Dead, We're Just on a Budget!"
You might be thinking: Wait, isn't NASA still a thing? Oh, absolutely. But today, NASA is less "boldly going where no man has gone before" and more "borrowing Elon's car to get there."

Here's the deal:
1. NASA outsourced the grunt work. Why build your own rockets when you can hire SpaceX for a fraction of the cost? It's like ordering Uber Eats instead of cooking—except the delivery guy is a Falcon 9, and your meal is a satellite.

2. The International Space Station (ISS) became a timeshare. NASA now splits the bill with Russia, Japan, and literally anyone who Venmos them $5.

3. Moon 2.0. The Artemis program aims to return humans to the moon by 2025. Key difference? This time, it's sponsored by Amazon Prime. (Just kidding. Probably.)

Why Capitalism Loves a Vacuum (The Space Kind, Not Your Dyson)
Space is the ultimate free-market playground: no environmental regulations, no minimum wage, and—until recently—no pesky laws about who owns a asteroid. It's the Wild West, but with more math.

Take Peter Beck, CEO of Rocket Lab, who compared launching satellites to "mowing the lawn in space." Or AstroForge, a startup that wants to mine asteroids because "Earth's resources are so 20th century." Meanwhile, lawyers are drafting intergalactic zoning laws, because nothing says "freedom" like a HOA on Mars.

The Absurdity Scale: Cold War vs. Space Capitalism
To recap the vibes:

Cold War Space Race	*2020s Space Capitalism*
- "For all mankind!".	- "For all my shareholders!".
- Astronauts as heroes.	- CEOs as "visionaries".
- Secret missile bases.	- Public rocket launch TikToks.
- Competing with Soviets.	- Competing with your own tax evasion.
- National pride.	- Personal brand deals.

Closing Thought: Is This Progress or a Midlife Crisis?
Let's be real: privatizing space is either humanity's greatest leap or a $300 billion midlife crisis for men who've run out of Earthly toys. But hey, at least we'll finally get that Mars Starbucks everyone's been asking for.

Next Chapter Preview: "Asteroid Mining: Because Who Wouldn't Trust a Billionaire to Exploit Space Rocks?"

Footnotes (Because Even Comedy Needs Citations)
1. Fun fact: Sputnik's beeps were just radio signals, but conspiracy theorists swear it was playing The Soviet National Anthem on loop.

2. Jeff Bezos's 4-minute space flight cost more per minute than Taylor Swift's Eras Tour tickets. Let that sink in.

3. Elon Musk once sold $1 billion in Tesla stock to fund SpaceX. When asked why, he said, "To make life multiplanetary. Duh."

Chapter 2: Asteroid Mining — The Ultimate Get-Rich-Quick Scheme (Until Space Pirates Show Up)

The New Gold Rush (But with More Zero-G Toilets)
Let's kick things off with a history lesson, because nothing says "comedy" like juxtaposing the California Gold Rush with space rocks. In 1849, thousands of folks raced west with pickaxes and dreams. In 2049, it'll be billionaires racing upward with robots and lawsuits. Why? Because asteroids are basically celestial piggy banks, stuffed with platinum, gold, and enough rare metals to make Elon Musk's Twitter feed look reasonable.

But here's the twist: mining an asteroid isn't like digging in your backyard. It's more like trying to rob a bank while riding a unicycle on a tightrope over a volcano. In space. With robots.

Why Asteroids? (Or: "Space Rocks Are the New Bitcoin")
First, let's answer the big question: Why would anyone mine a floating space boulder?

1. They're Stupidly Valuable: A single asteroid like 16 Psyche contains enough heavy metal to crash Earth's economy harder than a Millennial's avocado toast habit. Estimates? Oh, just $10,000 quadrillion. That's right—quadrillion. A number so big it makes Jeff Bezos's net worth look like a coupon for 10% off at Denny's.

2. They're Not Guarded (Yet): Unlike Earth's dwindling resources, asteroids are up for grabs. No environmentalists, no zoning laws, just… space.

3. Bragging Rights: Imagine your LinkedIn headline: "Chief Asteroid Officer at SpaceGold™. Skills: Zero-G drilling, dodging space lawsuits."

The Tech: Robots, Lasers, and the Eternal Fear of a Glitch
So, how do you mine an asteroid? Let's break it down:

1. Step 1: Find a Rock (Preferably One That Won't Kill Us)
    - NASA's got a catalog of "near-Earth objects," which is a fancy term for "asteroids that might hug us too hard."
    - Pro tip: Avoid "Apophis." Sounds like a pharaoh's curse, and coincidentally, it's due to swing by Earth in 2029.

2. Step 2: Send a Robot (Because Humans Are Terrible at Math)
    - Modern mining involves space drones with drills, lasers, and the existential dread of being 200 million miles from IT support.
    - Key feature: "Autonomous operation" (translation: "Hope it doesn't develop sentience and unionize").

3. Step 3: Haul It Home (Or to the Nearest Space Walmart)
    - Use a space tug (a glorified tow truck) to drag your asteroid into Earth's orbit. Risks include:
    - Accidentally creating a new moon.

- Explaining to the UN why your asteroid now blocks sunlight to Canada.

The Legal Black Hole: Who Owns a Space Rock?
Here's where things get spicy. The Outer Space Treaty of 1967 says no nation can own celestial bodies. But it says nothing about Elon Musk.

- The Loophole: Companies argue that while they can't own an asteroid, they can mine it. It's like saying, "I don't own the ocean, but I'll take this whale."

- The Lawyers (Yes, Space Lawyers): A new breed of attorneys now specialize in intergalactic bird law. Recent cases include "SpaceY vs. BlueOrigins: Who Left Their Satellite in My Orbit?"

- Satirical Quote: "If you can't afford a space lawyer, you'll be representing yourself… in a space court. Good luck with that."

Economic Fallout: When Platinum Becomes the New Plastic
Imagine this: You've successfully mined an asteroid and brought back 500 tons of platinum. Congrats! You've also just made Earth's platinum as valuable as a used Kleenex.

- Commodity Markets: "Asteroid metals could drop prices by 99%," say experts. Translation: Dentists who invested in SpaceGold™ are now crying into their diamond-encrusted floss.

- The Silver Lining: At least your toaster will finally have that platinum finish you've always wanted.

Space Pirates: The Yelp Review You Didn't See Coming

Ah, pirates. The romantic scourge of the 17th century, now rebooted with jetpacks and plasma cutters. How will asteroid mining attract these cosmic bandits?

1. The Loot: Precious metals in transit are basically a giant "Rob Me" sign in neon space letters.

2. The Tech: Pirates could hijack space tugs, hack drones, or just throw a net over your asteroid. Yes, a net. In space.

3. The Aesthetic: Picture a pirate with an eyepatch, a parrot (cyborg, obviously), and a peg leg that's also a rocket launcher. "Arrr, prepare to be boarded… and/or yeeted into the sun!"

Hypothetical Yelp Review:

☆ 1/5 Stars — "Took 3 years to get my space platinum. Then pirates stole it. Customer service blamed 'solar winds.'"

Conclusion: So, Should You Quit Your Job and Mine Asteroids?

Short answer: No.

Long answer: Unless you've got a few billion dollars, a team of rocket scientists, and a willingness to be sued by every country on Earth (plus a few angry pirates), stick to your day job. But hey, if you're feeling lucky, here's your starter kit:
- 1 Robot Miner (with extended warranty)
- 1 Space Lawyer (retainer: 10,000 Bitcoin)
- 1 Anti-Piracy Defense System (basically a giant space flyswatter)

Next Chapter Preview: "Lunar Real Estate: Because Who Wouldn't Want a Timeshare on a Desolate, Airless Rock?"

Footnotes (For the Pedantic and the Curious)

1. Fun fact: The first asteroid mining lawsuit was settled out of court for 100 tons of "future space gold." It's now framed in the CEO's bathroom.

2. The term "space tug" was almost called a "space Uber," but lawyers feared drivers would unionize.

3. Scientists confirm: A parrot can survive in space. But it'll cost you $5 million in oxygen fees.

Chapter 3: Lunar Real Estate — Buy Your Acre of Moon Dust Today! (Terms and Conditions Apply)

The Birth of Moon Scams: When Novelty Met Capitalism
Let's rewind to 1980. Disco was dead, shoulder pads were alive, and a man named Dennis Hope had a lightbulb moment: "What if I sold deeds to the moon?" Thus, the Lunar Embassy was born—a company that's been hustling faux lunar land titles for 40 years. For just $19.99, you could "own" a piece of the Sea of Tranquility. Terms and conditions? Buried in font size 2, like "This deed is worth less than the paper it's printed on… and also, the moon isn't ours to sell."

But here's the kicker: Hope isn't a billionaire. He's just a guy who read the 1967 Outer Space Treaty, noticed it didn't explicitly ban individuals from claiming celestial property, and went full Wolf of Wall Crater. To date, he's "sold" over 600 million acres. That's enough moon to give every Earthling a backyard… if only we could breathe there.

Moon Zoning Laws: Where Space Lawyers Earn Their Weight in Gold (Which They'll Mine from Asteroids)
Fast-forward to today. Companies like Lunar Land and Moon Estates are still peddling deeds, but now they've got competition: billionaires with rockets. Suddenly, the moon isn't just a romanticized rock—it's a fixer-upper planetoid ripe for gentrification.

But before you build your lunar McMansion, you'll need to navigate intergalactic zoning laws:
- The Sea of Tranquility: Zoned for "historic preservation" (i.e., no casinos near Neil Armstrong's footprints).
- Shackleton Crater: Prime location for solar panels, because it's one of the few spots with near-constant sunlight. Also, great for vampires.
- Mare Imbrium: The "bad part of the moon." Mostly lava plains and a suspicious number of crashed satellites.

And don't forget the HOA (Helium-3 Owners Association), which will fine you for:
- Leaving your lunar rover parked on the regolith.
- Failing to dust your solar panels.
- Hosting a rave in your biodome without a permit.

Elon's Moonbase Airbnb: Luxury Living (If You Ignore the Lack of Air)
Enter SpaceX's latest venture: Moonbase Alpha-1, a "luxury habitat" with "stunning Earthrise views" and "optional oxygen subscriptions." Features include:
- Zero-G Hot Tubs: Because nothing says relaxation like floating in a bubble of water that could freeze or boil depending on the sun.
- Artisanal Moon Cheese: Aged in a radiation-proof fridge.
- Wi-Fi: $10,000 per megabyte. Buffering Netflix? That'll be your life savings.

Mock Airbnb Listing:

"Cozy 1-bedroom crater. Sleeps 2 (if you spoon). Includes 'authentic' moon dust carpeting. No pets (unless they're photosynthetic). 5-star review if you survive the trip!"

The Fine Print: Why Your Deed is Worth Less Than a Used Spacesuit

Let's get real. That "lunar deed" you bought online? It's about as legally binding as a Trump University diploma. Here's why:

1. The Outer Space Treaty: Nations can't claim the moon, but it's silent on whether Jeff from Nebraska can. Spoiler: Jeff can't.
2. The "First Come, First Served" Fallacy: If SpaceX lands on your "property," Elon's lawyers will argue "possession is 9/10ths of space law."
3. The "Who's Gonna Stop Me?" Defense: Try evicting a billionaire from your moon plot. You'll need a rocket, a lawsuit, and a death wish.

Satirical Quote from a "Moon Realtor":

"This prime lunar acre has unlimited potential! Farming? Sure, if you can grow potatoes in a vacuum. Vacation home? Absolutely, if you don't mind a 3-day rocket ride. Investment opportunity? Just ignore the UN laughing in the background."

Lunar Timeshares: Because One Week a Year on a Barren Rock is Plenty

Why stop at deeds? Enter moon timeshares—the perfect gift for the relative you barely tolerate.

- Basic Package: One week in a pressurized pod. Bring your own oxygen.
- Deluxe Package: Includes a guided tour of Tycho Crater and a souvenir moon rock (Warning: May be a painted pebble).
- Platinum Package: Dinner with a "former astronaut" (actor in a SpaceX costume).

Infomercial Pitch:
"Call now, and we'll DOUBLE your offer! That's two weeks on the moon for the price of one! (Restrictions apply. Weeks non-consecutive. Must occur between 2075 and 2076. Oxygen sold separately.)"

The Future: Moon Condos, Evictions, and Space Squatters
By 2050, the moon will be a glorified gated community for the ultra-rich. Picture:
- Bezos's Blue Origin Estates: Gated craters with Amazon Prime delivery drones (5-7 business light-years).
- Branson's Virgin Lunar Lounge: A nightclub where the cover charge is a kidney.
- Musk's Marsgate Community: For people who bought moon land but got scammed into "upgrading" to a Martian dust bowl.

And when the first lunar squatters arrive? They'll be greeted by space lawyers serving eviction notices via laser pointer.

Conclusion: So, Should You Buy Lunar Land?
Only if you enjoy:

- Paying real money for imaginary dirt.
- Explaining to your grandkids why they'll never inherit your "moon fortune."
- Being the punchline of a future South Park episode.

But hey, at least you'll get a cool NFT deed… and a lifetime supply of regret.

Next Chapter Preview: "Interplanetary Trade: Why Your Amazon Package Will Take 300 Years (and Cost a Planet)"

Footnotes (For Those Who Skipped to the End)
1. Dennis Hope, the Lunar Embassy founder, once tried to sell Martian land too. The Mars reviews are in: "0/5 stars. No atmosphere. Literally."
2. The moon's Mare Imbrium translates to "Sea of Showers." Fitting, since buying land there will make you cry.
3. Elon's moonbase plans include a "Tesla Lunar Rover." Features: Autopilot (avoids craters 60% of the time), $20,000 "Ludicrous Speed" upgrade, and a cupholder that only fits a S'well bottle.

Chapter 4: Interplanetary Trade — Why Your Amazon Package Will Take 300 Years (and Cost a Planet)

The Dawn of the Space Trucker (Heroes with Wrenches and Zero Patience)

Let's start with the unsung heroes of the cosmos: space truckers. These are the folks hauling freeze-dried coffee from Martian Starbucks to the icy meth addicts on Pluto. Picture a grizzled pilot named Deke "Two-Ton" Johnson, sipping asteroid-belt bourbon in his cockpit, muttering about how "back in my day, we delivered to one moon and liked it."

The Job Perks:
- Zero-G Coffee Breaks: Where spilling your drink creates a floating brown nebula.
- Scenic Routes: Flying past Saturn's rings while your cargo of Venusian hot sauce slowly freezes.
- Cowboy Ethics: "You ain't a real trucker 'til you've outrun a pirate gang in the Kuiper Belt."

The Reality:
- Paycheck: Mostly space bucks (converted to Dogecoin at a 1:10,000,000 rate).
- Benefits: "Health insurance" is a first-aid kit duct-taped to a broken airlock.

Cryptocurrency in Space: Dogecoin to the Moon (Literally)

Why Dogecoin? Because Elon Musk once tweeted a meme, and now it's the official currency of Mars. Imagine trying to buy a moon rock with Bitcoin, only to realize the transaction takes 20 minutes due to light-speed lag.

How It Works (Sort Of):
- Mars Colony LLC pays miners in Dogecoin to maintain the blockchain.
- Miners set up rigs on Pluto for "cooling efficiency." (Spoiler: They freeze solid.)
- Everyone ignores the fact that one solar flare could erase the entire economy.

Satirical Quote from a Crypto Bro:
- "NFTs? Pfft. I'm investing in interstellar NFTs. This JPEG of Neptune's dark spot will make me rich… in 500 years."

Space Customs: Where Bureaucracy Meets Black Holes

Welcome to Galactic Customs, where the motto is "Your crap, our rules." Here's what happens when your shipment of Earth olives arrives at Mars:

- Tariffs: 300% tax because olives are "non-essential luxuries." (Meanwhile, Martian red sand is duty-free… for now.)
- Prohibited Items:
- Venusian lava lamps ("fire hazard").
- Europa's ice wine ("might melt and flood the cargo bay").

- Sentient alien plants ("unless certified emotional support flora").

Hypothetical Customs Form:

☑ I hereby swear my space tomatoes are not a bioweapon.

☑ I acknowledge that Jupiter's storms may delay delivery (and also obliterate it).

Space Prime: Because Bezos Won't Stop Until He Sells Uranus

Amazon's new Space Prime service promises "lightning-fast" delivery across the solar system. Terms apply.

Membership Benefits:
- Free Delivery: On orders over $10 million (excludes black holes, war zones, and Titan's methane lakes).
- Prime Video: Streams only Plan 9 from Outer Space and Spaceballs on loop.
- Alexa in Space: "Sorry, I can't find oxygen refill stations in your area. Would you like to hear a fun fact about supernovas?"

Mock Customer Review:

☆ 1/5 — "Ordered a self-heating space burrito. Arrived cold 75 years later. My great-grandkids say it's 'vintage.'"

Interplanetary Trade Wars: When Saturn's Moons Throw Shade

The year is 2075. The Jovian Alliance (Jupiter's moons) slashes prices on rocket fuel, sparking the Great Gas Giant Trade War.

- Saturn Strikes Back: Dumping cheap, glittery ring particles into the market.
- Uranus's Move: Imposing a "gravity tax" on all shipments passing its… orbital vicinity.
- Earth's Response: Sending a UN diplomat who accidentally starts a peace treaty in Klingon.

Fake News Headline:

"Titan Embargo Escalates: Europa Threatens to Block All shipments of Artisanal Ice!"

The Fine Print: Why Your Grandkids Will Inherit Your Space Debts

Buying a Martian rug? Congrats! Here's what you're really signing up for:

- Shipping Time: 3-5 centuries (depending on solar winds and whether SpaceX remembers your address).
- Returns Policy: "Must be in original condition, with all atmospheric seals intact. No returns for 'changed my mind after dying of old age.'"
- Customer Service: A chatbot that replies, "Your call is important to us. Please hold for 10,000 years."

Conclusion: Should You Shop Interplanetary?

Only if:

- You're okay with your package arriving posthumously.

- You enjoy explaining to your accountant why you claimed a "Plutonian tax deduction."
- You've always dreamed of owning a souvenir T-shirt that says "I Spent a Fortune on Space Shipping and All I Got Was This Lousy Meteor."

Next Chapter Preview: "Space Tourism: Where Billionaires Go to Cry in Zero-G (And Charge You for the Tissue)"

Footnotes (For the Astronomically Curious)

1. Fun fact: The first space trucker union formed in 2065. Their demands: "Less cosmic radiation, more bathroom breaks."
2. Dogecoin's Mars exchange rate: 1 DOGE = 0.0000001 grams of oxygen. A real steal!
3. The Great Gas Giant Trade War ended when everyone realized Uranus was just giggling.

Chapter 5: Space Tourism — Where Billionaires Go to Escape Their Tax Bills

The Ultimate Midlife Crisis Purchase

Let's face it: Once you've bought your third yacht, your private island starts to feel… pedestrian. Enter space tourism—the pièce de résistance of billionaire ennui. For the low, low price of a small country's GDP, you too can float in zero gravity, vomit elegantly into a vacuum-sealed bag, and Instagram it with the caption "Living my best life! 🌎 ✨" (hashtag #BlessedByPhysics).

Jeff Bezos kicked things off in 2021 by riding a phallic rocket to the edge of space, waving like a dad at a Disneyland parade. Richard Branson followed, sipping champagne in Virgin Galactic's "Spaceship Two" (which looks suspiciously like a glorified paper airplane). Meanwhile, Elon Musk is busy selling tickets to Mars, because of course he wants to die on the Red Planet. It's the ultimate "hold my beer" flex.

The "Vomit Comet" Economy (Gravity: Optional)

Space tourism isn't just about views—it's about surviving the Gauntlet of Nausea. Most tourists spend their flight oscillating between awe and the urge to hurl, thanks to the "Vomit Comet" effect. Pro tip: Swallow that $1,000 pre-flight "anti-motion-sickness" pill before you remember it's just a Tic Tac with a markup.

The Economics of Upchucking:

- Pre-Flight Training: $500,000. Includes learning to pee in a tube and a PowerPoint titled "Why You Shouldn't Panic in a Vacuum."
- In-Flight Photos: $250,000 per shot. Smiles cost extra.
- Post-Flight Therapy: $1 million/hour. "Yes, Karen, space is lonely. Let's unpack that."

Hypothetical Yelp Review:

⭐ 2/5 — "The 'Earthy sunrise' view was just my burrito floating past the window. Would not recommend."

Space Hotels: Where Luxury Meets Existential Dread

Forget the Ritz-Carlton. The future of hospitality is Orbital Suites™, a "five-star" hotel where the stars are real, but the oxygen is rationed.

Amenities Include:
- Zero-G Spa: Get a massage from a robot that may or may not short-circuit and yeet you into the stratosphere.
- Gourmet Dining: Freeze-dried "beef" Wellington paired with Tang (vintage 1969).
- Earthrise Balcony: Stare at the planet you're actively ignoring while climate change burns Miami.

Mock TripAdvisor Listing:

"Loved the anti-radiation pajamas! But the 'all-you-can-breathe' air package was a scam. 3/5 stars — would suffocate again."

The Shark Tank Pitch from Hell

Imagine pitching space tourism to investors:

Entrepreneur: "So, you know how people love cruises? Let's do that… but in SPACE! Risks? Pfft. Just add waivers!"

Mr. Wonderful: "You're telling me customers pay millions to risk death in a tin can? I'M IN!"

Barbara: "What's the exit strategy?"

Entrepreneur: "Death by supernova. Duh."

The Dark Side of Space Selfies

Let's talk about the real cost of that Instagram moment:
- Carbon Footprint: One joyride emits more CO2 than a village does in a decade. But hey, #EcoWarrior, right?
- Space Junk: Your commemorative "I ♥ SPACE" keychain is now orbiting Earth at 17,000 mph. Congrats, you've littered on a cosmic scale.
- Morality: Explaining to your kids why you blew their college fund to take a Buzz Aldrin cosplay pic.

Satirical Quote from a Space Influencer:
"Ugh, Earth is so basic. Can't wait to monetize my zero-G skincare routine on TikTok!"

Conclusion: Is It Worth It?

If you've got cash to burn and a pathological need to outdo your Tesla-owning neighbor, absolutely. For everyone else? Stick to planet-based vacations. At least in Cancún, the margaritas don't cost $10,000—and the sea turtles won't judge you for destroying the ozone layer.

Next Chapter Preview: "The Dark Side of Space Capitalism: Polluting the Final Frontier (and Other Crimes Against the Galaxy)"

Footnotes (For the Discerning Space Cadet)

1. Fun fact: Jeff Bezos's spaceflight lasted 11 minutes—roughly the same attention span as a golden retriever.
2. Virgin Galactic's safety video is just 10 minutes of Richard Branson laughing. No refunds.
3. Elon's Mars tickets include a "survival kit": a protein bar, a flashlight, and a PDF titled "How to Mine Water or Die Trying."

Chapter 6: The Dark Side of Space Capitalism — Pollution, Plutocrats, and the Probability of War Over Uranus

Introduction: The Great Space Screw-Up

Welcome to the chapter where we ask: "Is space capitalism just Earth's bad habits with worse WiFi?" Spoiler: Yes. From orbiting junkyards to billionaire ego wars, humanity's quest to conquer the cosmos is less Star Trek and more Dumb and Dumber in Zero-G. Let's dive into the dumpster fire we've lit among the stars.

1. Space Pollution: Earth's Orbit is Now a Flying Landfill

Imagine Earth's orbit as your college dorm room after a three-day rave: cluttered, sticky, and filled with regret. Thanks to 150 million pieces of space junk, our planet is wrapped in a shrapnel blanket traveling at 17,500 mph.

- The Culprits: Dead satellites, lost screwdrivers, and Elon's Tesla Roadster (still blasting "Space Oddity" to no one).
- The "Cleanup": Startups like JunkBusters Inc. promise to vacuum debris with giant space nets. Reviews? "☆ 2/5 — Caught one satellite and three Starbucks cups. Charged me $10 million. Would not trash again."
- Future Outlook: By 2050, Earth's orbit will resemble Mad Max, but with more Wi-Fi routers.

Satirical Corporate Slogan:

"SpaceX: Making Mars Habitable… After We've Made Earth Uninhabitable!"

2. Plutocrats in Space: Because One Planet Wasn't Enough to Ruin

Meet the Galactic 1%—billionaires who've swapped yachts for rockets and tax havens for literal moons.

- Elon's Mars Colony: A utopia where workers mine water ice for $3/hour and pay rent in Dogecoin. Benefits include:
- Free SpaceX merch (if you survive 10 years).
- A complimentary "How Not to Die" PDF.
- Bezos's Orbital Mansions: Zero-G penthouses with "Earthrise views" and a strict no-poors policy.
- Branson's Space Casino: Where the house always wins… because the house is in space and you can't leave.

Hypothetical Headline:

"Billionaires Launch 'Save the Earth' Charity—From Their Private Moon Bases. (Tax-Deductible, Of Course.)"

3. War Over Uranus: The Interplanetary Misunderstanding Heard 'Round the Galaxy.

It started with a typo. In 2075, a Martian mining corp misspelled "Your Anus" in a treaty, sparking the Great Uranus War.

- The Conflict: Earth-based conglomerates vs. the Jovian Alliance over who "owns" Uranus's methane. (Spoiler: No one. But try telling that to a lawyer in a spacesuit.)
- Weaponry:
- Satellite Lasers: For slicing enemy drones and toasting space bagels.
- Asteroid Nukes: Because nothing says "diplomacy" like hurling a space rock.
- The Peace Treaty: Signed at McDonald's Lunar Franchise #69. Terms included free fries for life and a ban on pronouncing it "Uranus."

Mock UN Resolution:
"Article 1: No fighting in the asteroid belt. Article 2: Seriously, stop laughing at Uranus."

4. The Ethical Black Hole: Are We the Baddies?
Let's face it: Space capitalism has the moral depth of a petri dish.

- Robot Labor Rights: Mars robots unionize, demanding better pay and less lava. Management responds: "Resistance is futile. Also, your warranty's void."
- Alien Encounters: When we finally meet ET, he'll sue us for copyright infringement ("You stole our probe design!") and demand royalties on asteroid metals.
- Space Colonialism: Replicating Earth's worst traditions—but with more lasers.

Satirical Quote from a Space CEO:

"Sustainability is key. That's why we're strip-mining asteroids and charging extra for biodegradable spacesuits."

Conclusion: So… Is Space Capitalism a Giant Leap or a Faceplant?
In the end, space capitalism is like a toddler with a flamethrower: ambitious, dangerous, and doomed to burn down the sandbox. Will we learn? Probably not. But hey, at least our great-great-grandkids will have front-row seats to the galaxy's first trash-powered supernova.

Footnotes (For the Astronomically Jaded)
1. Fun fact: Cleaning up space junk costs more than the GDP of Luxembourg. Good thing Luxembourg owns half the junk.
2. The Great Uranus War ended when someone misheard "Ceasefire" as "Cheese fries."
3. Elon's Mars colony handbook includes a chapter titled "Crying in a Pressure Suit: A How-To."

www.ingramcontent.com/pod-product-compliance
Lightning Source LLC
Chambersburg PA
CBHW060034040426
42333CB00042B/2443